Birds of Prey

EAGLES

LYDIA LUKIDIS

BLACK RABBIT BOOKS

Bolt is published by Black Rabbit Books
P.O. Box 227, Mankato, Minnesota, 56002
www.blackrabbitbooks.com

Marysa Storm, editor; Michael Sellner, designer and photo researcher

Names: Lukidis, Lydia, author.
Title: Eagles / by Lydia Lukidis.
Description: Mankato, Minnesota : Black Rabbit Books, [2023] | Series: Bolt. birds of prey | Includes bibliographical references and index. | Audience: Ages 8-12 | Audience: Grades 4-6 | Summary: "Equipped with strong senses and powerful features, eagles rule the skies. They also capture the imagination of readers everywhere. With an engaging design focused around expertly leveled text and colorful infographics that support visual literacy, introduce reluctant and beginning readers to the life cycle, features, and diets of these powerful birds of prey" – Provided by publisher.
Identifiers: LCCN 2020017999 (print) | LCCN 2020018000 (ebook) | ISBN 9781623105525 (hardcover) | ISBN 9781644664834 (paperback) | ISBN 9781623105587 (ebook)
Subjects: LCSH: Hawks–Juvenile literature.
Classification: LCC QL696.F32 L843 2023 (print) | LCC QL696.F32 (ebook) | DDC 598.9/44–dc23
LC record available at https://lccn.loc.gov/2020017999
LC ebook record available at https://lccn.loc.gov/2020018000

Image Credits

Alamy: Arterra Picture Library, 18; blickwinkel, 12–13; Nature Picture Library, 21; REUTERS, 4–5; Rocky Grimes, 22; Dreamstime: Craigrjd, 25; Kevin Gower, 25; Pascal Halder, 15; Sergey Uryadnikov, 8–9; Wirestock, 28–29; Getty: Mark Newman, 21; Shutterstock: a_v_d, 25; basel101658, 7; Charlotte Heijmans, 11; DenisKrivoy, 7; FloridaStock, 14; FotoIdee, 15; Henk Bogaard, 32; Independent birds, 25; Jesus Giraldo Gutierrez, 11; Jiri Hrebicek, 18; Lana Langlois, 25; Matt_Williams, 3; Michal Lukaszewicz, 26–27; PROMA1, 11; rbrown10, 6–7; Rupert Wilkey, Cover; Sergey Uryadnikov, 16–17; Smit, 25; Susan Schmitz, 25; Viju Jose, 1, 31; Viktorya170377, 7; Vladimirs Suscinskis, 23; Wang LiQiang, 11; Zheltyshev, 25

Contents

CHAPTER 1

FEARSOME Fliers

An eagle soars through the air. It stares at the ground below, looking for its next meal.

The large bird spots a rabbit. It waits for the right moment. Then it swoops down. The race is on! The rabbit speeds across the ground. But the little **mammal** isn't fast enough. The eagle's sharp talons sink into its **prey**. Time for lunch!

COMPARING WINGSPANS

Strong Birds

Eagles are powerful birds of prey. These **predators** eat only meat. They have long, mighty wings and curved beaks. They also have big claws called talons. These talons are hooked. Eagles use them to grab and kill prey.

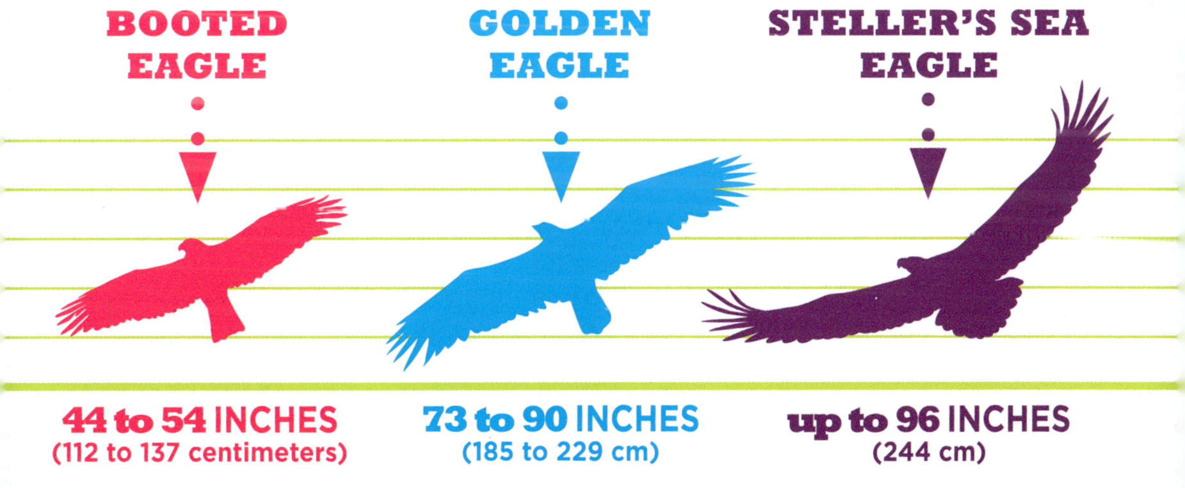

PARTS OF AN EAGLE

TALONS
CURVED BEAK

Different Types

There are about 60 types of eagles. Scientists often divide them into four main groups. Each type is a bit different. Booted eagles have feathers on their legs. Fish eagles have rough toes that help them hold slippery fish. Snake eagles are small to medium-sized. They mostly eat snakes. Harpy eagles are quite large.

BOOTED EAGLES

FISH EAGLES

SNAKE EAGLES

HARPY EAGLES

When eagles are full, they keep eating food. This food doesn't go into their stomachs, though. Birds store extra food in their **crops**. The extra food helps them survive if they can't find new food.

Huge Eyes and Tiny Stomachs

All eagles have large eyes and amazing eyesight. Their eyesight is much stronger than a human's vision. Most can see a rabbit about 2 miles (3 kilometers) away.

Eagles are big birds, but their stomachs are tiny. A bald eagle's stomach is only about the size of a walnut. An eagle can still eat up to a third of its body weight, though.

CHAPTER 2

At HOME and on the Hunt

Eagles live and hunt around the world. They rule the skies in many different **habitats**. They live in hot and cold places.

Eagle Habitats

GRASSLANDS

FORESTS

MOUNTAINS

EAGLE RANGE MAP

Eagles live on every continent except Antarctica.

ASIA
EUROPE
AFRICA
AUSTRALIA
ANTARCTICA

HUNTING

STEALING FOOD

Finding Food

Some eagles hunt from a perch, such as a tree branch. Others look for meals while flying through the sky. They all swoop down to surprise their prey.

Eagles often eat fish, snakes, other birds, and small mammals. Large harpy eagles hunt bigger prey, such as monkeys. Eagles sometimes steal other animals' food too.

CHAPTER 3

EAGLE Families

Eagles often fly alone, but they do meet to **mate**. Eagles usually have the same partner each year. Most won't find a new mate unless the other dies.

Pairs build nests in trees, on cliffs, or other high places. They use twigs, branches, and grass to line the nests.

To bond, eagle pairs will fly up high and lock talons. They then fall to the ground doing cartwheel spins. They break apart at the last minute.

Eagle parents must bring food to their babies. To feed their young, adults drop chunks of food in the babies' mouths.

Baby Eagles

After mating, female eagles lay one to four eggs. The parents then take turns sitting on the eggs to keep them warm. About six to eight weeks later, the eggs crack open.

Eagle chicks are helpless. They can't fly and must spend all their time in the nest. As they grow, they practice flapping their wings. By about 10 weeks old, they take their first uneasy flights.

CHAPTER 4

PREDATORS and Other Dangers

Adult eagles are strong and powerful. They are at the top of the food chain. Not many animals would dare attack them. However, some animals attack eagle nests. Eagle eggs and chicks make tasty meals for many mammals, **reptiles**, and other birds.

Eagle Food Chain

This food chain shows what eats baby eagles. It also shows what eagles eat.

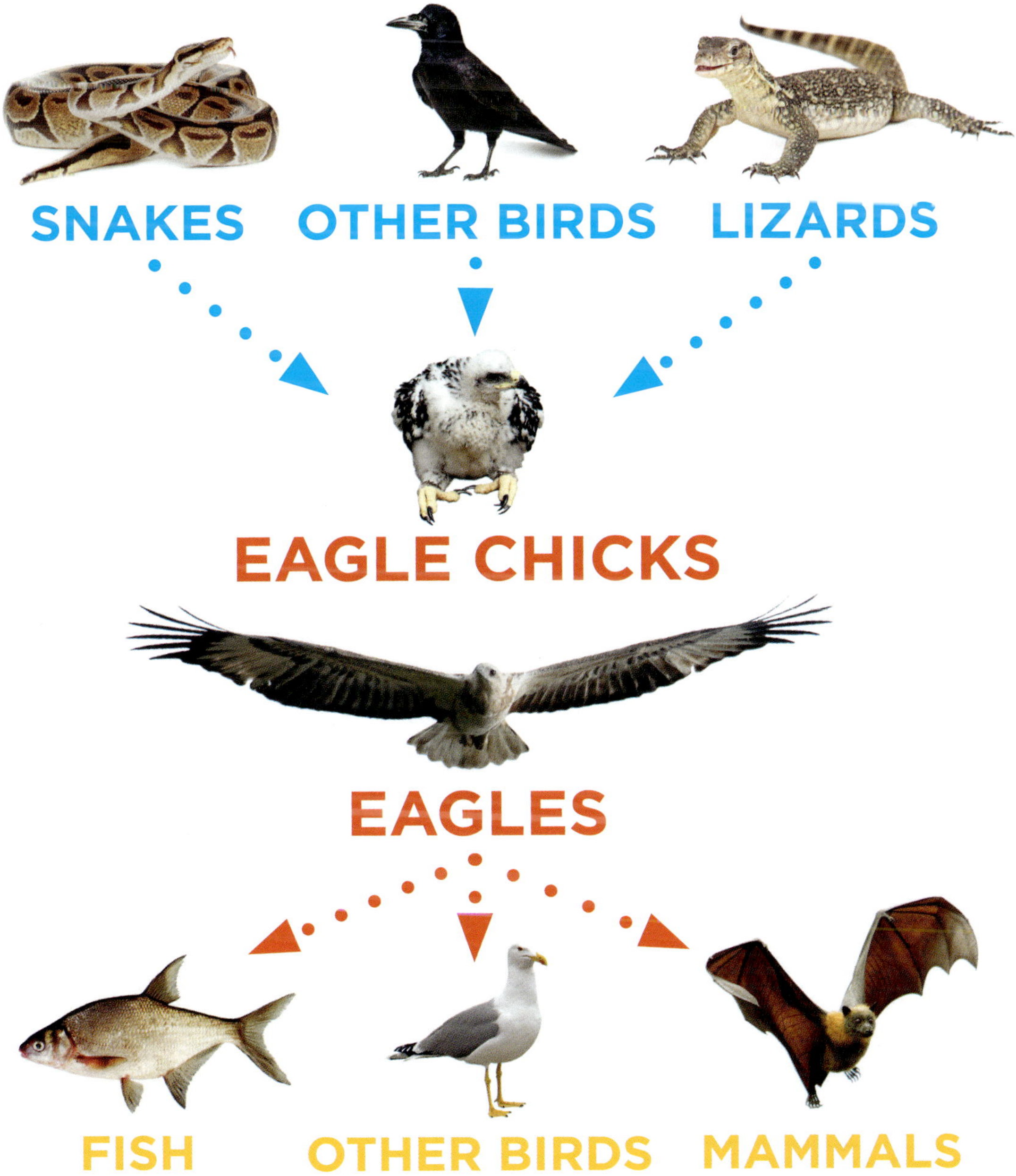

Disappearing

Humans are a threat to eagles. Some people think eagles are a danger to their pets or farm animals. They shoot the eagles. People also hurt eagles by destroying their habitats. The birds have less space to live and hunt.

Some eagles are now **endangered**. They might disappear completely. Eagles play an important role in the food chain. Humans must make sure these fearsome fliers don't vanish.

ENDANGERED
PHILIPPINE EAGLE

By the NUMBERS

about **2** POUNDS (1 kilogram)

HOW MUCH BOOTED EAGLES WEIGH

20 FEET (6 METERS)

depth of largest recorded bald eagle nest

about
120,000
NUMBER OF
GOLDEN EAGLES
IN THE WORLD
10 to 30
YEARS
LIFE SPAN

GLOSSARY

crop (KRAHP)—an enlargement just above the stomach of a bird or insect where food is temporarily stored

endangered (in-DAYN-jurd)—close to no longer existing

habitat (HAB-uh-tat)—the place where a plant or animal grows or lives

mammal (MAH-muhl)—a type of animal that feeds milk to its young and usually has hair or fur

mate (MAYT)—to join together to produce young; it also means one of a pair that joins together to produce young.

predator (PRED-uh-tuhr)—an animal that eats other animals

prey (PRAY)—an animal hunted or killed for food

reptile (REP-tahyl)—a cold-blooded animal that breathes air and has a backbone; most reptiles lay eggs and have scaly skin.

LEARN MORE

BOOKS

Clasky, Leonard. *The Bald Eagle.* Return from Extinction. New York: Gareth Stevens Publishing, 2023.

Grodzicki, Jenna. *Baby Eagles.* Animal Babies. Minneapolis: Bearport Publishing Company, 2022.

McDonald, Amy. *Eagles.* Animals in My Yard. Minneapolis: Bellwether Media, 2022.

WEBSITES

African Fish Eagle
www.natgeokids.com/au/primary-resource/african-fish-eagle-primary-resource/

Fun Eagle Facts for Kids
www.sciencekids.co.nz/sciencefacts/animals/eagle.html

Harpy Eagle
sdzwildlifeexplorers.org/animals/harpy-eagle

INDEX